The ESSENTIALS of

Geometry I

**Staff of
Research & Education Association**

Edited by
Mark Shapiro, M.A.
Adjunct Professor of Mathematics
University of Hartford
West Hartford, CT

Research & Education Association
Visit our website at
www.rea.com

Y0-BZE-801

Research & Education Association
61 Ethel Road West
Piscataway, New Jersey 08854
E-mail: info@rea.com

THE ESSENTIALS®
OF GEOMETRY I

Year 2006 Printing

Copyright © 2002, 2000, 1998, 1996, 1987
by Research & Education Association, Inc.
All rights reserved. No part of this book
may be reproduced in any form without
permission of the publisher.

Printed in the United States of America

Library of Congress Control Number 2001091421

International Standard Book Number 0-87891-606-7

ESSENTIALS® and REA® are registered trademarks of
Research & Education Association, Inc.

What REA's Essentials® Will Do for You

This book is part of REA's celebrated *Essentials*® series of review and study guides, relied on by tens of thousands of students over the years for being complete yet concise.

Here you'll find a summary of the very material you're most likely to need for exams, not to mention homework—eliminating the need to read and review many pages of textbook and class notes.

This slim volume condenses the vast amount of detail characteristic of the subject matter and summarizes the **essentials** of the field. The book provides quick access to the important facts, principles, theorems, concepts, and equations in the field.

It will save you hours of study and preparation time.

This *Essentials*® book has been prepared by experts in the field and has been carefully reviewed to ensure its accuracy and maximum usefulness. We believe you'll find it a valuable, handy addition to your library.

Larry B. Kling
Chief Editor

CONTENTS

CHAPTER 1

Method of Proof

1.1 Logic

Definition 1

A statement is a sentence that is either true or false, but not both.

Definition 2

If *a* and *b* are statements, then a statement of the form "*a* and *b*" is called the conjunction of *a* and *b*, denoted by $a \wedge b$.

Definition 3

The disjunction of two statements *a* and *b* is shown by the compound statement "*a* or *b*," denoted by $a \vee b$.

Definition 4

The negation of a statement *q* is the statement "not *q*," denoted by $\sim q$.

Definition 5

The compound statement "if a, then b," denoted by $a \rightarrow b$, is called a condition statement or an implication.

"If a" is called the hypothesis or premise of the implication; "then b" is called the conclusion of the implication.

Further, statement a is called the antecedent of the implication, and statement b is called the consequent of the implication.

Definition 6

The converse of $a \rightarrow b$ is $b \rightarrow a$.

Definition 7

The contrapositive of $a \rightarrow b$ is $\sim b \rightarrow \sim a$.

Definition 8

The inverse of $a \rightarrow b$ is $\sim a \rightarrow \sim b$.

Definition 9

The statement of the form "p if and only if q," denoted by $p \leftrightarrow q$, is called a biconditional statement.

Definition 10

An argument is valid if the truth of the premises means that the conclusion must also be true.

Definition 11

Intuition is the process of making generalizations based on insight.

2

Basic Principles, Laws, and Theorems

1. Any statement is either true or false (The Law of the Excluded Middle).

2. A statement cannot be both true and false (The Law of Contradiction).

3. The converse of a true statement is not necessarily true.

4. The converse of a definition is always true.

5. For a theorem to be true, it must be true for all cases.

6. A statement is false if one false instance of the statement exists.

7. The inverse of a true statement is not necessarily true.

8. The contrapositive of a true statement is true and the contrapositive of a false statement is false.

9. If the converse of a true statement is true, then the inverse is true. Likewise, if the converse is false, the inverse is false.

10. Statements that are either both true or false are said to be logically equivalent.

11. If a given statement and its converse are both true, then the conditions in the hypothesis of the statement are both necessary and sufficient for the conclusion of the statement.

    If a given statement is true but its converse is false, then the conditions are sufficient but not necessary for the conclusion of the statement.

    If a given statement and its converse are both false, then the conditions are neither sufficient nor necessary for the statement's conclusion.

1.2　Deductive Reasoning

An arrangement of statements that would allow you to deduce the third one from the preceding two is called a syllogism. A syllogism has three parts:

The first part is a general statement concerning a whole group. This is called the major premise.

The second part is a specific statement which indicates that a certain individual is a member of that group. This is called the minor premise.

The last part of a syllogism is a statement to the effect that the general statement which applies to the group also applies to the individual. This third statement of a syllogism is called a deduction.

EXAMPLE A: Properly Deduced Argument

A)　Major Premise: All birds have feathers.

B)　Minor Premise: An eagle is a bird.

C)　Deduction: An eagle has feathers.

The technique of employing a syllogism to arrive at a conclusion is called deductive reasoning.

If a major premise which is true is followed by an appropriate minor premise which is true, a conclusion can be deduced which must be true, and the reasoning is valid. However, if a major premise which is true is followed by an inappropriate minor premise which is also true, a conclusion cannot be deduced.

EXAMPLE B: Improperly Deduced Argument

A)　Major Premise: All people who vote are at least 18 years old.

B)　Improper Minor Premise: Jane is at least 18.

C)　Illogical Deduction: Jane votes.

The flaw in example B is that the major premise stated in A makes a condition on people who vote, not on a person's age. If statements B and C are interchanged, the resulting three-part deduction would be logical.

1.3   Indirect Proof

Indirect proofs involve considering two possible outcomes—the result we would like to prove and its negative—and then showing, under the given hypothesis, that a contradiction of prior known theorems, postulates, or definitions is reached when the negative is assumed.

Postulate 1

A proposition contradicting a true proposition is false.

Postulate 2

If one of a given set of propositions must be true, and all except one of those propositions have been proved to be false, then this one remaining proposition must be true.

The method of indirect proof may be summarized as follows:

Step 1.   List all the possible conclusions.

Step 2.   Prove all but one of those possible conclusions to be false (use Postulate 1 given).

Step 3.   The only remaining possible conclusion is proved true according to Postulate 2.

EXAMPLE

When attempting to prove that in a scalene triangle the bisector of an angle cannot be perpendicular to the opposite side, one method of solution could be to consider the two possible conclusions:

1)   the bisector can be perpendicular to the opposite side, or

5

2) the bisector cannot be perpendicular to the opposite side.

Obviously one and only one of these conclusions can be true; therefore, if we can prove that all of the possibilities, except one, are false, then the remaining possibility must be a valid conclusion. In this example, it can be proven that for all cases, the statement which asserts that the bisector of an angle of a scalene triangle can be perpendicular to the opposite side is false. Therefore, the contradicting possibility—that the bisector cannot be perpendicular to the opposite side—is in fact true.

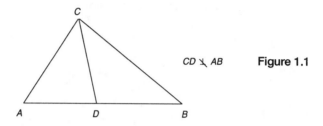

$CD \perp AB$ **Figure 1.1**

1.4   Inductive Reasoning

Definition 1

Inductive reasoning is a method of reasoning in which one draws conclusions or generalizations from several known particular cases. The resulting conclusion is called an induction.

Definition 2

Inductive reasoning is the drawing of a conclusion based on experimenting with particular examples. A typical mathematical induction is the Method of Proof whereby:

A) the conditions of the statement are valid for the smallest possible value of n;

B) the conditions are assumed for the general case, say $n = k$;

C) the conditions are tested and verified for $n = k + 1$.

If A is proved true, and C is true when B is assumed, then the statement is valid for all n greater than or equal to the smallest n.

1.5 Defined and Undefined Terms: Axioms, Postulates, and Assumptions; Theorems and Corollaries

To build a logical system of mathematics, the first step is to take a known and then move to what is not known. The terms which we will accept as known are called undefined terms. We accept certain basic terms as undefined, since their definition would of necessity include other undefined terms. Examples of some important undefined terms with characteristics that you must know are:

A) Set: The sets we will be concerned with will have clearly defined characteristics.

B) Point: Although we represent points on paper with small dots, a point has no size, thickness, or width. A point is denoted by a capital letter.

C) Line: A line is a series of adjacent points which extends indefinitely. A line can be either curved or straight; however, unless otherwise stated, the term "line" refers to a straight line. A line is denoted by listing two points on the line and drawing a line with arrows on top, i.e., $\overleftrightarrow{AB}$.

D) Plane: A plane is the collection of all points lying on a flat surface which extends indefinitely in all directions. Imagine holding a record cover in a room, and imagine that the record cover divides the entire room. Remember that a plane has no thickness.

We use these undefined terms to construct defined terms so we can describe more sophisticated expressions.

Necessary characteristics of a good definition are:

A) It names the term being defined.

B) It uses only known terms or accepted undefined terms.

7

C) It places the term into the smallest set to which it belongs.

D) It states the characteristics of the defined term which distinguish it from the other members of the set.

E) It contains the least possible amount of information.

F) It is always reversible.

Axioms, postulates, and assumptions are the statements in geometry which are accepted as true without proof, whereas theorems are the statements in geometry which are proven to be true.

A corollary is a theorem that can be deduced easily from another theorem or from a postulate.

In this text the term postulate is used exclusively, instead of axiom or assumption.

Postulate 1

A quantity is equal to itself (reflexive law).

Postulate 2

If two quantities are equal to the same quantity, they are equal to each other (transitive law).

Postulate 3

If a & b are any quantities, and $a = b$, then $b = a$ (symmetric law).

Postulate 4

The whole is equal to the sum of its parts.

Postulate 5

If equal quantities are added to equal quantities, the sums are equal quantities.

Postulate 6

If equal quantities are subtracted from equal quantities, the differences are equal quantities.

Postulate 7

If equal quantities are multiplied by equal quantities, the products are equal quantities.

Postulate 8

If equal quantities are divided by equal quantities (not 0), the quotients are equal quantities.

CHAPTER 2

Points, Lines, Planes, and Angles

Definition 1

If A and B are two points on a line, then the line segment AB is the set of points on that line between A and B and including A and B, which are called the endpoints. The line segment is referred to as $\overline{AB}$.

$A \qquad B$

Figure 2.1

Definition 2

A half-line is the set of all the points on a line on the same side of a dividing point, not including the dividing point, denoted by $\overrightarrow{AB}$.

$A \qquad B$

Figure 2.2

Definition 3

Let A be a dividing point on a line. Then a ray is the set of all the points on a half-line and the dividing point itself. The dividing point is called the endpoint or the vertex of the ray. The ray AB shown below is denoted by $\overrightarrow{AB}$.

Figure 2.3

Definition 4

Three or more points are said to be collinear if and only if they lie on the same line.

Definition 5

Let X, Y, and Z be three collinear points. If Y is between X and Z, then $\overrightarrow{YX}$ and $\overrightarrow{YZ}$ are called opposite rays.

Figure 2.4

Definition 6

The absolute value of x, denoted by $|x|$, is defined as

$$|x| = \begin{cases} x & \text{if} \quad x > 0 \\ 0 & \text{if} \quad x = 0 \\ -x & \text{if} \quad x < 0 \end{cases}$$

Definition 7

The absolute value of the difference of the coordinates of any two points on the real number line is the distance between those two points.

Definition 8

The length of a line segment is the distance between its endpoints.

11

Definition 9

Congruent segments are segments that have the same length.

Definition 10

The midpoint of a segment is defined as the point of the segment which divides the segment into two congruent segments. (The midpoint is said to bisect the segment.)

Definition 11

The bisector of a line segment is a line that divides the line segment into two congruent segments.

Definition 12

An angle is a collection of points which is the union of two rays having the same endpoint. An angle such as the one illustrated in figure 2.5 can be referred to in any of the following ways:

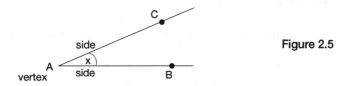

Figure 2.5

A) by a capital letter which names its vertex, i.e., ∡A;

B) by a lowercase letter or number placed inside the angle, i.e., ∡x;

C) by three capital letters, where the middle letter is the vertex and the other two letters are not on the same ray, i.e., ∡ CAB or ∡BAC, both of which represent the angle illustrated in the figure.

Definition 13

A set of points is coplanar if all the points lie in the same plane.

Definition 14

Two angles with a common vertex and a common side, but no common interior points, are called adjacent angles.

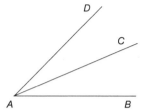

Figure 2.6

In the above figure, ⦨ *DAC* and ⦨ *BAC* are adjacent angles; ⦨ *DAB* and ⦨ *BAC* are not adjacent angles.

Definition 15

Vertical angles are two angles with a common vertex and with sides that are two pairs of opposite rays.

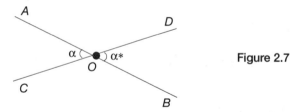

Figure 2.7

(⦨α and ⦨α* are vertical angles.)

Definition 16

An acute angle is an angle whose measure is larger than 0° but smaller than 90°.

Definition 17

An angle whose measure is 90° is called a right angle.

13

Definition 18

An obtuse angle is an angle whose measure is larger than 90° but less than 180°.

Definition 19

An angle whose measure is 180° is called a straight angle. Note: Such an angle is, in fact, a straight line.

Definition 20

An angle whose measure is greater than 180° but less than 360° is called a reflex angle.

Definition 21

Complementary angles are two angles, the sum of whose measures equals 90°.

Definition 22

Supplementary angles are two angles, the sum of whose measures equals 180°.

Definition 23

Congruent angles are angles of equal measure.

Definition 24

A ray bisects (is the bisector of) an angle if the ray divides the angle into two angles that have equal measure.

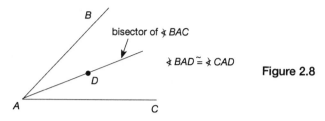

bisector of ∡ BAC

∡ BAD ≅ ∡ CAD

Figure 2.8

Definition 25

If the two non-common sides of adjacent angles form opposite rays, then the angles are called a linear pair. Note that α and β are supplementary.

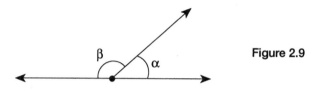

Figure 2.9

Definition 26

Two lines are said to be perpendicular if they intersect and form right angles. The symbol for perpendicular (or, is perpendicular to) is ⊥ ; $\overleftrightarrow{AB}$ is perpendicular to $\overleftrightarrow{CD}$ is written $\overleftrightarrow{AB} \perp \overleftrightarrow{CD}$.

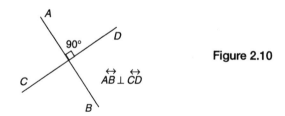

Figure 2.10

Definition 27

A line, a ray, or a line segment which bisects a line segment and is also perpendicular to that segment is called a perpendicular bisector of the line segment.

Definition 28

The distance from a point to a line is the measure of the perpendicular line segment from the point to that line. Note: This is the shortest possible distance of the point to the line, which is later stated in Chapter 7, Theorem 1.

Definition 29

Two or more distinct lines are said to be parallel (∥) if and only if they are coplanar and they do not intersect.

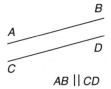

Figure 2.11

AB ∥ CD

Definition 30

The projection of a given point on a given line is the foot of the perpendicular drawn from the given point to the given line.

The foot of a perpendicular from a point to a line is the point where the perpendicular meets the line.

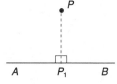

Figure 2.12

P_1 is the projection of P on $\overleftrightarrow{AB}$

Definition 31

The projection of a segment on a given line (when the segment is not perpendicular to the line) is a segment with endpoints that are the projections of the endpoints of the given line segment onto the given line.

16

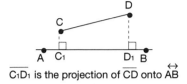

Figure 2.13

C_1D_1 is the projection of $\overline{CD}$ onto $\overleftrightarrow{AB}$

Postulate 1 (The Point Uniqueness Postulate)

Let n be any positive number, then there exists exactly one point N of $\overrightarrow{AB}$ such that $AN = n$. (AN is the length of $\overline{AN}$.)

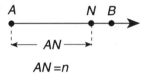

Figure 2.14

$AN = n$

Postulate 2 (The Line Postulate)

Any two distinct points determine one and only one line which contains both points.

Figure 2.15

Postulate 3 (The Point Betweenness Postulate)

Let A and B be any two points. Then there exists at least one point (and in fact an infinite number of such points) of $\overleftrightarrow{AB}$ such that P is between A and B, with $AP + PB = AB$.

Figure 2.16

17

Postulate 4

Two distinct straight lines can intersect at most at only one point.

Figure 2.17

Postulate 5

The shortest line between any two points is a straight line.

Figure 2.18

Postulate 6

There is a one-to-one correspondence between the real numbers and the points of a line. That is, to every real number there corresponds exactly one point of the line, and to every point of the line there corresponds exactly one real number. (In other words, a line has an infinite number of points between any two distinct points.)

Postulate 7

One and only one perpendicular can be drawn to a given line through any point on that line. Given point O on line $\overleftrightarrow{AB}$, $\overleftrightarrow{OC}$ represents the only perpendicular to $\overleftrightarrow{AB}$ which passes through O.

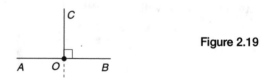

Figure 2.19

Postulate 8

The perpendicular bisector of a line segment is unique.

Postulate 9 (The Plane Postulate)

Any three non-collinear points determine one and only one plane that contains those three points.

Postulate 10 (The Points-in-a-Plane Postulate)

If two distinct points of a line lie in a given plane, then the line lies in that plane.

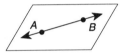

Figure 2.20

Postulate 11 (Plane Separation Postulate)

Any line in a plane separates the plane into two half planes.

Postulate 12

Given an angle, there exists one and only one real number between 0 and 180 corresponding to it. Note: $m \angle A$ refers to the measurement of angle A.

Postulate 13 (The Angle Sum Postulate)

If A is in the interior of $\angle XYZ$, then $m\angle XYZ = m\angle XYA + m\angle AYZ$.

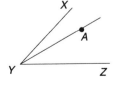

Figure 2.21

19

Postulate 14 (The Angle Difference Postulate)

If *P* is in the exterior of ⊀*ABC* and in the same half-plane (created by edge *BC*) as *A*, then *m* ⊀*ABP* = *m* ⊀*PBC* – *m* ⊀*ABC*.

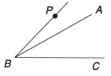

Figure 2.22

Theorem 1

All right angles are equal.

Theorem 2

All straight angles are equal.

Theorem 3

Supplements of the same or equal angles are themselves equal.

Theorem 4

Complements of the same or equal angles are themselves equal.

Theorem 5

Vertical angles are equal.

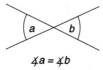

Figure 2.23

⊀*a* = ⊀*b*

20

Theorem 6

Two supplementary angles are right angles if they have the same measure.

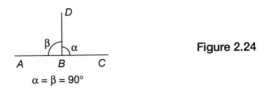

Figure 2.24

$\alpha = \beta = 90°$

Theorem 7

If two lines intersect and form one right angle, then the lines form four right angles.

Theorem 8

Any point on the perpendicular bisector of a given line segment is equidistant from the ends of the segment.

Figure 2.25

Theorem 9

If a point is equidistant from the ends of a line segment, this point must lie on the perpendicular bisector of the segment.

Theorem 10

If two points are equidistant from the ends of a line segment, these points determine the perpendicular bisector of the segment.

Theorem 11

Every line segment has exactly one midpoint.

Theorem 12

There exists one and only one perpendicular to a line through a point outside the line.

Take point P outside line $\overleftrightarrow{AB}$. $\overleftrightarrow{OP}$ represents the only perpendicular to $\overleftrightarrow{AB}$ which passes through P.

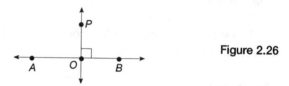

Figure 2.26

Theorem 13

If the exterior sides of adjacent angles are perpendicular to each other, then the adjacent angles are complementary.

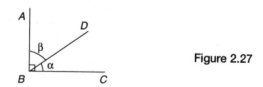

Figure 2.27

Theorem 14

Adjacent angles are supplementary if their exterior sides form a straight line.

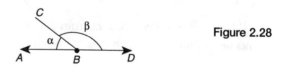

Figure 2.28

22

CHAPTER 3

Congruent Angles and Congruent Line Segments

Definition 1

Two or more geometric figures are congruent when they have the same shape and size. The symbol for congruence is $\cong$; hence, if triangle *ABC* is congruent to triangle *DEF*, we write $\triangle ABC \cong \triangle DEF$.

Definition 2

Two line segments are congruent if and only if they have the same measure.

Note: The expression "if and only if" can be used any time both a statement and the converse of that statement are true. Using definition 2 we can rewrite the statement as two line segments have the same measure if and only if they are congruent. The two statements are identical.

23

Definition 3

Two angles are congruent if and only if they have the same measure.

Theorem 1

Every line segment is congruent to itself.

Theorem 2

Every angle is congruent to itself.

Let R be a relation on a set A. Then:

R is reflexive if aRa for every a in A.

R is symmetric if aRb implies bRa.

R is antisymmetric if aRb and bRa imply $a = b$.

R is transitive if aRb and bRc imply aRc.

Note: The term aRa means the relation R performed on a yields a. The term aRb means the relation R performed on a yields b.

By definition, a relation R is called an equivalence relation if relation R is reflexive, symmetric, and transitive.

Postulate 1

Congruence of segments is an equivalence relation.

(1) Congruence of segments is reflexive.

If $\overline{AB} \cong \overline{AB}$, $\overline{AB}$ is congruent to itself.

(2) Congruence of segments is symmetric.

If $\overline{AB} \cong \overline{CD}$, then $\overline{CD} \cong \overline{AB}$.

(3) Congruence of segments is transitive.

If $\overline{AB} \cong \overline{CD}$ and $\overline{CD} \cong \overline{EF}$, then $\overline{AB} \cong \overline{EF}$.

24

Postulate 2

Congruence of angles is an equivalence relation, i.e., reflexive, symmetric, and transitive.

Postulate 3

Any geometric figure is congruent to itself.

Postulate 4

A congruence may be reversed.

Postulate 5

Two geometric figures congruent to the same geometric figure are congruent to each other.

Theorem 3

Given a line segment $\overline{AB}$ and a ray $\overrightarrow{XY}$, there exists one and only one point O on $\overrightarrow{XY}$ such that $\overline{AB} \cong \overline{XO}$.

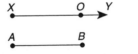

Figure 3.1

Theorem 4

If $AB = CD$, Q bisects $\overline{AB}$ and P bisects $\overline{CD}$, then $\overline{AQ} \cong \overline{CP}$, $AQ = CP$.

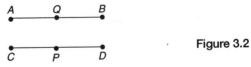

Figure 3.2

25

Theorem 5

If $m \sphericalangle ABC = m \sphericalangle DEF$, and $\overrightarrow{BX}$ and $\overrightarrow{EY}$ bisect $\sphericalangle ABC$ and $\sphericalangle DEF$, respectively, then $m \sphericalangle ABX = m \sphericalangle DEY$.

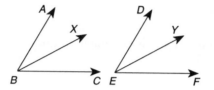

Figure 3.3

Theorem 6

Let P be in the interior of $\sphericalangle ABC$ and Q be in the interior of $\sphericalangle DEF$. If $m \sphericalangle ABP = m \sphericalangle DEQ$ and $m \sphericalangle PBC = m \sphericalangle QEF$, then $m \sphericalangle ABC = m \sphericalangle DEF$.

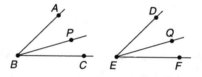

Figure 3.4

Theorem 7

Let P be in the interior of $\sphericalangle XYZ$ and Q be in the interior of $\sphericalangle ABC$. If $m \sphericalangle XYZ = m \sphericalangle ABC$ and $m \sphericalangle XYP = m \sphericalangle ABQ$ then $m \sphericalangle PYZ = m \sphericalangle QBC$.

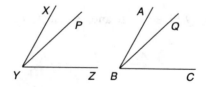

Figure 3.5

26

CHAPTER 4

Triangles and Congruent Triangles

4.1   Triangles

Definition 1

A closed three-sided geometric figure is called a triangle. The points of the intersection of the sides of a triangle are called the vertices of the triangle.

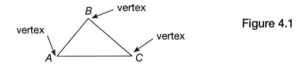

Figure 4.1

Definition 2

The perimeter of a triangle is the sum of the measures of the sides of the triangle.

Definition 3

A triangle with no equal sides is called a scalene triangle.

Figure 4.2

Definition 4

A triangle having two equal sides is called an isosceles triangle. The third side is called the base of the triangle. The angle opposite the base is called the vertex angle.

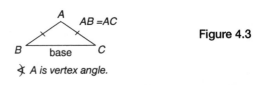

Figure 4.3

Definition 5

A side of a triangle is a line segment whose endpoints are the vertices of two angles of the triangle.

Definition 6

An interior angle of a triangle is an angle formed by two sides and includes the third side within its collection of points.

Definition 7

An equilateral triangle is a triangle having three equal sides.

Figure 4.4

Definition 8

A triangle with one obtuse angle (greater than 90°) is called an obtuse triangle.

Figure 4.5

Definition 9

An acute triangle is a triangle with three acute angles (less than 90°).

Figure 4.6

Definition 10

A triangle with a right angle is called a right triangle. The side opposite the right angle in a right triangle is called the hypotenuse of the right triangle. The other two sides are called legs of the right triangle.

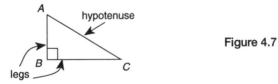

Figure 4.7

Definition 11

An altitude of a triangle is a line segment from a vertex of the triangle perpendicular to the opposite side.

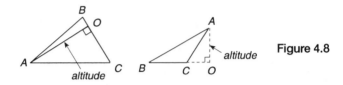

Figure 4.8

Definition 12

A line segment connecting a vertex of a triangle and the midpoint of the opposite side is called a median of the triangle.

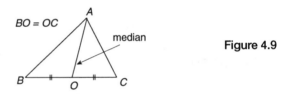

Figure 4.9

Definition 13

A line that bisects and is perpendicular to a side of a triangle is called a perpendicular bisector of that side.

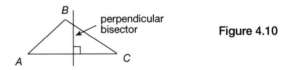

Figure 4.10

Definition 14

An angle bisector of a triangle is a line that bisects an angle and extends to the opposite side of the triangle.

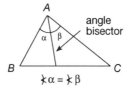

Figure 4.11

Definition 15

The line segment that joins the midpoints of two sides of a triangle is called a midline of the triangle.

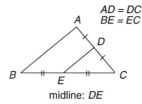

Figure 4.12

Definition 16

An exterior angle of a triangle is an angle formed outside a triangle by one side of the triangle and the extension of an adjacent side.

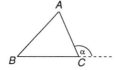

Figure 4.13

Definition 17

A triangle whose three interior angles have equal measure is said to be equiangular.

Definition 18

Three or more lines (or rays or segments) are concurrent if there exists one point common to all of them, that is, if they all intersect at the same point.

Theorem 1

The three lines containing the altitudes of a triangle are concurrent.

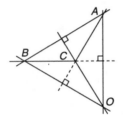

Figure 4.14

Theorem 2

The medians of a triangle are concurrent at a point which is two-thirds the distance from any vertex to the midpoint of the opposite side.

Point O is called the centroid of $\triangle ABC$.

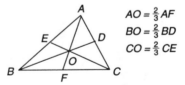

$AO = \frac{2}{3} AF$
$BO = \frac{2}{3} BD$
$CO = \frac{2}{3} CE$

Figure 4.15

Theorem 3

The perpendicular bisectors of the sides of a triangle are concurrent at a point that is equidistant from any vertex of the triangle.

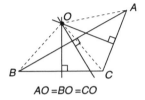

Figure 4.16

$$AO = BO = CO$$

Theorem 4

The angle bisectors of a triangle are concurrent at a point which is equidistant from any side of the triangle.

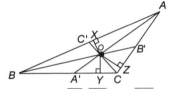

Figure 4.17

Angle bisectors $\overline{AA'}$, $\overline{BB'}$, and $\overline{CC'}$ meet at point O, and $OX = OY = OZ$.

Theorem 5

The measure of an exterior angle of a triangle is equal to the sum of the measures of the two nonadjacent interior angles of that triangle

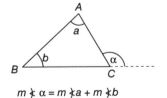

Figure 4.18

$$m \angle \alpha = m \angle a + m \angle b$$

33

Theorem 6

Every angle of a triangle has one and only one bisector.

Theorem 7

The midline of a triangle is parallel to the third side of the triangle.

Theorem 8

The midline of a triangle is half as long as the third side of the triangle.

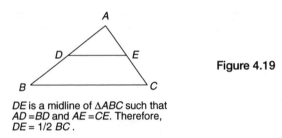

Figure 4.19

DE is a midline of $\triangle ABC$ such that
$AD = BD$ and $AE = CE$. Therefore,
$DE = 1/2\ BC$.

Theorem 9

The sum of the measures of the interior angles of a triangle is 180°.

Theorem 10

If two angles of one triangle are equal respectively to two angles of a second triangle, their third angles are equal.

Theorem 11

A triangle can have at most one right or obtuse angle.

34

Theorem 12

If a triangle has two equal angles, then the sides opposite those angles are equal.

Theorem 13

If two sides of a triangle are equal, then the angles opposite those sides are equal.

Theorem 14

The sum of the exterior angles of a triangle, taking one angle at each vertex, is 360°.

Theorem 15

A line that bisects one side of a triangle, and is parallel to a second side, bisects the third side.

Theorem 16 (The Law of Cosines)

For a given triangle $\triangle ABC$ with sides of length a, b, and c,

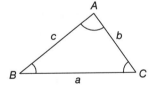

Figure 4.20

$$a^2 = b^2 + c^2 - 2bc\cos A$$

4.2 Isosceles, Equilateral, and Right Triangles

Theorem 1

The length of the median to the hypotenuse of a right triangle is equal to one-half the length of the hypotenuse.

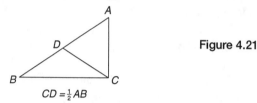

Figure 4.21

$$CD = \tfrac{1}{2}AB$$

Theorem 2

In a right triangle, the square of the hypotenuse is equal to the sum of the squares of the other two sides. This is commonly known as the theorem of Pythagoras or the Pythagorean theorem.

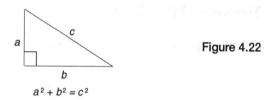

Figure 4.22

$$a^2 + b^2 = c^2$$

Theorem 3

If a triangle has sides of length a, b and c, and $c^2 = a^2 + b^2$, then the triangle is a right triangle.

Theorem 4

In a 30°–60° right triangle, the hypotenuse is twice the length of the side opposite the 30° angle. The side opposite the 60° angle is equal to the length of the side opposite the 30° angle multiplied by $\sqrt{3}$.

In an isosceles 45° right triangle, the hypotenuse is equal to the length of one of its legs multiplied by $\sqrt{2}$.

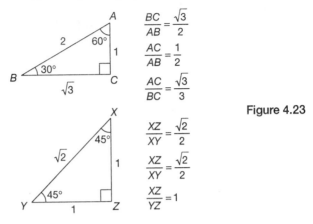

$$\frac{BC}{AB} = \frac{\sqrt{3}}{2}$$

$$\frac{AC}{AB} = \frac{1}{2}$$

$$\frac{AC}{BC} = \frac{\sqrt{3}}{3}$$

Figure 4.23

$$\frac{XZ}{XY} = \frac{\sqrt{2}}{2}$$

$$\frac{XZ}{XY} = \frac{\sqrt{2}}{2}$$

$$\frac{XZ}{YZ} = 1$$

Theorem 5

The altitude of an equilateral triangle equals $\dfrac{\sqrt{3}}{2}$ times the measure of a side of the triangle.

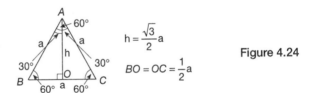

$$h = \frac{\sqrt{3}}{2}a$$

$$BO = OC = \frac{1}{2}a$$

Figure 4.24

Theorem 6

The base angles of an isosceles triangle are congruent, where the two sides adjacent to the base are equal.

Corollary 1

The angle bisectors of the base angles of an isosceles triangle are congruent.

Corollary 2

The bisector of the vertex angle of an isosceles triangle is also the perpendicular bisector of the base of the triangle.

Corollary 3

All equiangular triangles are also equilateral and every equilateral triangle is equiangular.

Corollary 4

The three angles of an equilateral triangle each have a measure of 60°.

Corollary 5

The acute angles of a right triangle are complementary.

Corollary 6

The two acute angles of an isosceles right triangle each have a measure of 45°.

4.3   Congruent Triangles

Definition

Two polygons are congruent if there is a one-to-one correspondence between their vertices such that all pairs of corresponding sides have equal measures and all pairs of corresponding angles have equal measures. This is denoted by $\cong$.

Theorem 1

Triangle congruence is an equivalence relation.

38

By definition, a relation R is called an equivalence relation if relation R is reflexive, symmetric, and transitive.

Properties of Congruence:

A) Reflexive property: $\triangle ABC \cong \triangle ABC$.

B) Symmetric property: If $\triangle ABC \cong \triangle DEF$, then $\triangle DEF \cong \triangle ABC$.

C) Transitive Property: If $\triangle ABC \cong \triangle DEF$, and $\triangle DEF \cong \triangle RST$, then $\triangle ABC \cong \triangle RST$.

Postulate 1

If three sides of one triangle are equal, respectively, to three sides of a second triangle, the triangles are congruent (SSS = SSS).

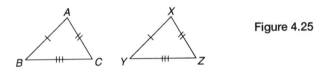

Figure 4.25

Postulate 2

If two sides and the included angle of one triangle are equal, respectively, to two sides and the included angle of a second triangle, the triangles are congruent (SAS = SAS).

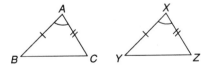

Figure 4.26

39

Postulate 3

If two angles and the included side of one triangle are equal, respectively, to two angles and the included side of a second triangle, the triangles are congruent (ASA = ASA).

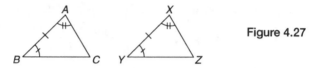

Figure 4.27

Theorem 2

If two angles and a not-included side of one triangle are equal, respectively, to two angles and a not-included side of a second triangle, the triangles are congruent (*AAS = AAS*).

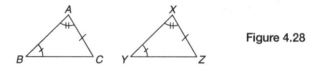

Figure 4.28

Theorem 3

Corresponding parts of congruent triangles are equal.

Theorem 4

Two right triangles are congruent if the two legs of one right triangle are congruent to the two corresponding legs of the other right triangle.

Theorem 5

If the hypotenuse and an acute angle of one right triangle are equal, respectively, to the hypotenuse and an acute angle of a second right triangle, the triangles are congruent.

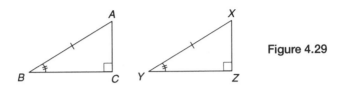

Figure 4.29

Theorem 6

If the hypotenuse and a leg of one right triangle are equal, respectively, to the hypotenuse and a leg of a second right triangle, the right triangles are congruent.

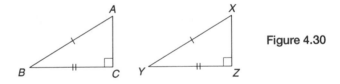

Figure 4.30

Theorem 7

If a leg and the adjacent acute angle of one right triangle are congruent, respectively, to a leg and the adjacent acute angle of another right triangle, then these two right triangles are congruent.

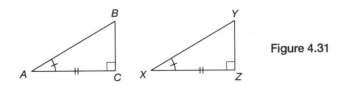

Figure 4.31

41

4.4 Areas of Triangles

Theorem 1

The area of a triangle is given by the formula $A = 1/2bh$, where b is the length of a base and h is the corresponding height of the triangle.

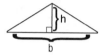

Figure 4.32

Corollary

If two triangles are congruent, they have the same area.

Theorem 2

The area of a triangle equals one-half the product of any two adjacent sides and the sine of their included angle.

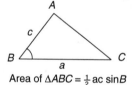

Area of $\triangle ABC = \frac{1}{2} ac \sin B$

Figure 4.33

Theorem 3

Two triangles with bases of equal length, and altitudes to their bases of equal length, have equal areas.

Theorem 4

Triangles that share the same base and have their vertices on a line parallel to the base, have equal areas.

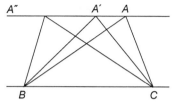

Figure 4.34

Area of $\triangle A''BC$ = Area of $\triangle A'BC$ = Area of $\triangle ABC$

Theorem 5

In a given triangle, the product of the length of any side and the length of the altitude drawn to that side is equal to the product of the length of any other side and the length of the altitude drawn to that side.

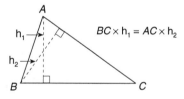

$BC \times h_1 = AC \times h_2$

Figure 4.35

Theorem 6

The area of a right triangle is equal to one-half the product of the lengths of its two legs.

Theorem 7

A median drawn to a side of a triangle divides the triangle into two triangles that are equal in area.

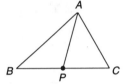

Figure 4.36

If $BP = PC$, then Area of $\triangle ABP$ = Area of $\triangle APC$

Theorem 8

The area of a triangle, with sides of lengths a, b, and c, is given by the formula

$$A = \sqrt{s(s-a)(s-b)(s-c)}$$

where $s = 1/2(a+b+c)$, the semi-perimeter of the triangle. The above formula is commonly referred to as Heron's formula or Hero's formula.

Theorem 9

The area of an equilateral triangle is given by the formula

$$A = \frac{x^2\sqrt{3}}{4}$$

where x is the length of a side of the triangle.

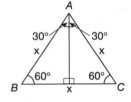

Figure 4.37

Theorem 10

The areas of two triangles with equal bases have the same ratio as the ratio of their altitudes; and the areas of two triangles with equal altitudes have the same ratio as the ratio of their bases.

Theorem 11

The altitude of an equilateral triangle equals $\dfrac{\sqrt{3}}{2}$ times the length of a side of the triangle.

Theorem 12

The area of an equilateral triangle equals $\dfrac{1}{\sqrt{3}}$ times the square of the length of an altitude of the triangle.

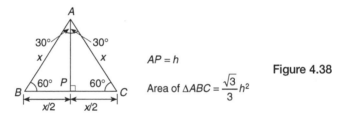

$AP = h$

Area of $\triangle ABC = \dfrac{\sqrt{3}}{3}h^2$

Figure 4.38

Theorem 13

The area of an isosceles triangle with congruent sides length l and an included angle of measure a is: $A = 1/2\ l^2 \sin a$. Area is also given by the formula

$$A = h^2 \tan \tfrac{\alpha}{2};$$

h is the length of the altitude to the side opposite to the angle α.

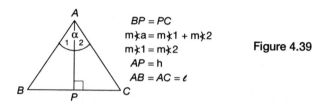

$BP = PC$
$m \not{\star} a = m \not{\star} 1 + m \not{\star} 2$
$m \not{\star} 1 = m \not{\star} 2$
$AP = h$
$AB = AC = \ell$

Figure 4.39

CHAPTER 5

Parallelism

Definition 1

A transversal of two or more lines is a line that cuts across these lines in two or more points, one point for each line.

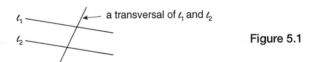

a transversal of ℓ_1 and ℓ_2

Figure 5.1

Definition 2

If two lines are cut by a transversal, nonadjacent angles on opposite sides of the transversal but on the interior of the two lines are called alternate interior angles.

$\not{\kern-1pt}\alpha$ and $\not{\kern-1pt}\beta$ are alternate interior angles

Figure 5.2

Definition 3

If two lines are cut by a transversal, nonadjacent angles on opposite sides of the transversal and on the exterior of the two lines are called alternate exterior angles.

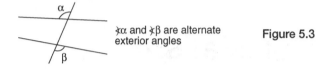

∡α and ∡β are alternate exterior angles

Figure 5.3

Definition 4

If two lines are cut by a transversal, angles on the same side of the transversal and in corresponding positions with respect to the lines are called corresponding angles.

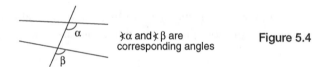

∡α and ∡β are corresponding angles

Figure 5.4

Definition 5

Two lines are called parallel lines if and only if they are in the same plane (coplanar) and do not intersect. The symbol for parallel, or is parallel to, is ‖: $\overleftrightarrow{AB}$ is parallel to $\overleftrightarrow{CD}$ is written $\overleftrightarrow{AB}$ ‖ $\overleftrightarrow{CD}$.

Definition 6

The distance between two parallel lines is the length of the perpendicular segment from any point on one line to the other line.

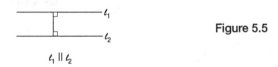

ℓ_1 ‖ ℓ_2

Figure 5.5

Postulate 1

Given a line ℓ and a point P not on line ℓ, there is one and only one line through point P that is parallel to line ℓ.

Postulate 2

Two coplanar lines are either intersecting lines or parallel lines.

Postulate 3

If two (or more) lines are perpendicular to the same line, then they are parallel to each other.

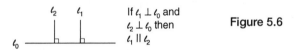

If $\ell_1 \perp \ell_0$ and $\ell_2 \perp \ell_0$ then $\ell_1 \parallel \ell_2$

Figure 5.6

Postulate 4

If two lines are cut by a transversal so that alternate interior angles are equal, the lines are parallel.

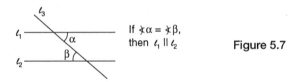

If $\sphericalangle\alpha = \sphericalangle\beta$, then $\ell_1 \parallel \ell_2$

Figure 5.7

Theorem 1

If two lines are parallel to the same line, then they are parallel to each other.

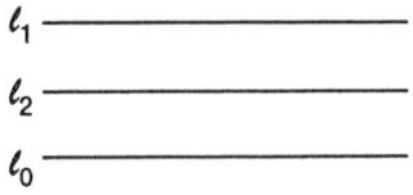

If $\ell_1 \parallel \ell_0$ and $\ell_2 \parallel \ell_0$, then $\ell_1 \parallel \ell_2$

Figure 5.8

49

Theorem 2

If a line is perpendicular to one of two parallel lines, then it is also perpendicular to the other line.

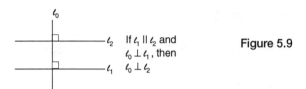

If $\ell_1 \parallel \ell_2$ and $\ell_0 \perp \ell_1$, then $\ell_0 \perp \ell_2$

Figure 5.9

Theorem 3

If two lines being cut by a transversal form congruent corresponding angles, then the two lines are parallel.

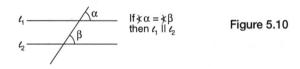

If $\sphericalangle \alpha = \sphericalangle \beta$ then $\ell_1 \parallel \ell_2$

Figure 5.10

Theorem 4

If two lines being cut by a transversal form interior angles on the same side of the transversal that are supplementary, then the two lines are parallel.

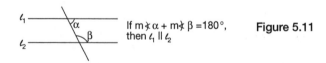

If $m\sphericalangle \alpha + m\sphericalangle \beta = 180°$, then $\ell_1 \parallel \ell_2$

Figure 5.11

Theorem 5

If a line is parallel to one of two parallel lines, it is also parallel to the other line.

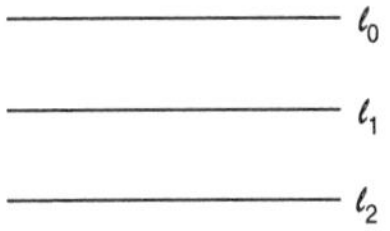

$$\ell_0$$
$$\ell_1$$
$$\ell_2$$

If $\ell_1 \parallel \ell_2$
and $\ell_0 \parallel \ell_1$ then
$\ell_0 \parallel \ell_2$

Figure 5.12

Theorem 6

If two parallel lines are cut by a transversal, then:

(A) The alternate interior angles are congruent.

(B) The corresponding angles are congruent.

(C) The consecutive interior angles are supplementary.

(D) The alternate exterior angles are congruent.

Theorem 7

Parallel lines are always the same distance apart.

Corollary 1

If a line intersects one of two parallel lines, it also intersects the other line.

Corollary 2

If two lines are cut by a transversal so that alternate interior angles are not equal, the lines are not parallel.

Corollary 3

If two lines are cut by a transversal so that corresponding angles are not equal, the lines are not parallel.

Corollary 4

If two lines are cut by a transversal so that two interior angles on the same side of the transversal are not supplementary, the lines are not parallel.

Corollary 5

If two nonparallel lines are cut by a transversal, the pairs of alternate interior angles are not equal.

Corollary 6

If line A is perpendicular to one of two parallel lines, and if another line B is perpendicular to the second of the two parallel lines, then lines A and B are parallel to each other.

Corollary 7

If three or more parallel lines intercept congruent segments on one transversal, then they intercept congruent segments on any transversal.

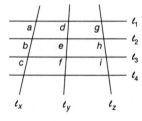

Figure 5.13

For example:

Given: Four parallel lines, ℓ_1, ℓ_2, ℓ_3 and ℓ_4, intersected by three transversals, ℓ_x, ℓ_y and ℓ_z, with the lengths of the transversals between parallel lines represented by the letters a through i as shown.

If $a = b = c$, then $d = e = f$ and $g = h = i$.

52

CHAPTER 6

Quadrilaterals

6.1 Parallelograms

Definition 1

A quadrilateral is a polygon with four sides.

Definition 2

A parallelogram is a quadrilateral whose opposite sides are parallel.

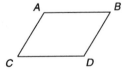

Figure 6.1

Definition 3

Two angles that have their vertices at the endpoints of the same side of a parallelogram are called consecutive angles.

Definition 4

The perpendicular segment connecting any point of a line containing one side of the parallelogram to the line containing the opposite side of the parallelogram is called an altitude of the parallelogram.

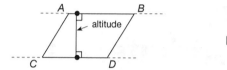

Figure 6.2

Definition 5

A diagonal of a polygon is a line segment joining any two non-consecutive vertices.

Theorem 1

A diagonal of a parallelogram divides the parallelogram into two congruent triangles.

Theorem 2

Consecutive angles of a parallelogram are supplementary.

Theorem 3

If both pairs of opposite sides of a quadrilateral are equal, then the quadrilateral is a parallelogram.

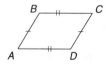

Figure 6.3

Theorem 4

The diagonals of a parallelogram bisect each other.

Theorem 5

If two opposite sides of a quadrilateral are both parallel and equal, the quadrilateral is a parallelogram.

$\overline{BC} \parallel \overline{AD}$ Figure 6.4

Theorem 6

If the diagonals of a quadrilateral bisect each other, then the quadrilateral is a parallelogram.

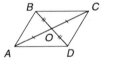

Figure 6.5

Theorem 7

Opposite sides of a parallelogram are equal.

Theorem 8

Nonconsecutive angles of a parallelogram are equal.

Theorem 9

If both pairs of opposite angles of a quadrilateral are congruent, the quadrilateral is a parallelogram.

Theorem 10

If one angle of a quadrilateral is congruent to the opposite angle, and one side is parallel to the opposite side, then the quadrilateral is a parallelogram.

$\overline{AB} \parallel \overline{CD}$

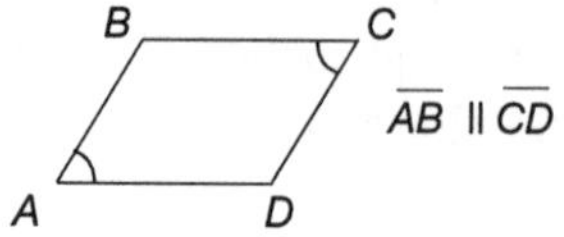

Figure 6.6

Theorem 11

If one angle of a quadrilateral is congruent to its opposite, and one side is congruent to its opposite, then the quadrilateral is a parallelogram.

Corollary

Segments of parallel lines intercepted between parallel lines are congruent.

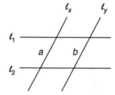

Figure 6.7

If $\ell_1 \parallel \ell_2$ and $\ell_x \parallel \ell_y$, then $a = b$

6.2 Rectangles

Definition

A rectangle is a parallelogram with one right angle.

Theorem 1

All angles of a rectangle are right angles.

Theorem 2

The diagonals of a rectangle are equal.

Theorem 3

If the diagonals of a parallelogram are equal, the parallelogram is a rectangle.

Theorem 4

If a quadrilateral has four right angles, then it is a rectangle.

Theorem 5

If a parallelogram is inscribed within a circle, then it is a rectangle.

Figure 6.8

In a circle O with inscribed quadrilateral $ABCD$, quadrilateral $ABCD$ is a parallelogram; therefore, quadrilateral $ABCD$ is a rectangle.

6.3 Rhombi

Definition

A rhombus is a parallelogram with two adjacent sides equal.

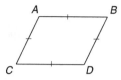

Figure 6.9

Theorem 1

All sides of a rhombus are equal.

Theorem 2

The diagonals of a rhombus are perpendicular to each other.

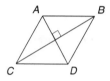

Figure 6.10

Theorem 3

The diagonals of a rhombus bisect the angles of the rhombus.

Theorem 4

If the diagonals of a parallelogram are perpendicular, the parallelogram is a rhombus.

Theorem 5

If a quadrilateral has four equal sides, then it is a rhombus.

Theorem 6

A parallelogram is a rhombus if either diagonal of the parallelogram bisects the angles of the vertices it joins.

58

6.4 Squares

Definition

A square is a rhombus with a right angle.

Figure 6.11

Theorem 1

A square is an equilateral quadrilateral.

Theorem 2

A square has all the properties of parallelograms and rectangles.

Theorem 3

A rhombus is a square if one of its interior angles is a right angle.

Theorem 4

In a square, the measure of either diagonal can be calculated by multiplying the length of any side by the square root of 2.

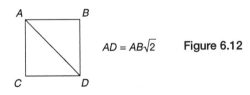

$AD = AB\sqrt{2}$ Figure 6.12

6.5 Trapezoids

Definition 1

A trapezoid is a quadrilateral with two and only two sides parallel. The parallel sides of a trapezoid are called bases.

Definition 2

The median of a trapezoid is the line joining the midpoints of the nonparallel sides.

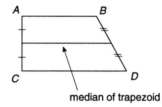

Figure 6.13

median of trapezoid

Definition 3

The perpendicular segment connecting any point in the line containing one base of the trapezoid to the line containing the other base is the altitude of the trapezoid.

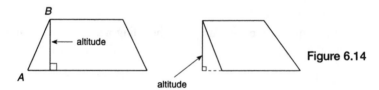

altitude

altitude

Figure 6.14

Definition 4

An isosceles trapezoid is a trapezoid whose nonparallel sides are equal. A pair of angles including only one of the parallel sides is called a pair of base angles.

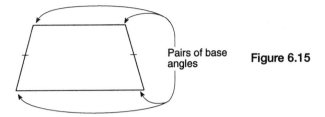

Pairs of base
angles

Figure 6.15

Theorem 1

The median of a trapezoid is parallel to the bases and equal to one-half their sum.

Theorem 2

The base angles of an isosceles trapezoid are equal.

Theorem 3

The diagonals of an isosceles trapezoid are equal.

Theorem 4

The opposite angles of an isosceles trapezoid are supplementary.

Theorem 5

If one pair of base angles of a trapezoid are congruent, then the trapezoid is isosceles.

Theorem 6

A trapezoid is isosceles if any angle and its opposite are supplementary.

Theorem 7

If the diagonals of a trapezoid are congruent, then the trapezoid is isosceles.

CHAPTER 7

Geometric Inequalities

7.1 Postulates and Theorems

Postulate 1

A quantity may be substituted for its equal in any inequality.

Postulate 2

A whole quantity is greater than any of its parts.

Postulate 3

The relation "$<$" is transitive; that is, if $a < b$ and $b < c$, then $a < c$.

Postulate 4

If the same quantity is added to both sides of an inequality, the sums are still unequal and in the same order.

Postulate 5

If equal quantities are added to unequal quantities, the sums are still unequal and in the same order.

(If $a < b$ and $c = d$, then $a + c < b + d$.)

Postulate 6

If unequal quantities are added to unequal quantities of the same order, the sums are unequal quantities and in the same order.

(If $a < b$ and $c < d$, then $a + c < b + d$.)

Postulate 7

If equal quantities are subtracted from unequal quantities, the differences are unequal and in the same order.

(If $b > c$, then $b - a > c - a$.)

Postulate 8

If unequal quantities are subtracted from equal quantities, the differences are unequal and in the opposite order.

(If $a < b$ and $c = d$, then $c - a > d - b$.)

Postulate 9

If both sides of an inequality are multiplied by a positive number, the products are unequal and in the same order.

(If $a < b$ and c is a positive number, then $ac < bc$.)

Postulate 10

If both sides of an inequality are multiplied by a negative number, the products are unequal in the opposite order.

(If $a < b$ and c is a negative number, then $ac > bc$.)

Postulate 11

If unequal quantities are divided by equal positive quantities, the quotients are unequal in the same order.

(If $a > b$, c and d are positive, and $c = d$, then $a \div c > b \div d$.)

Postulate 12

If unequal quantities are divided by equal negative quantities, the quotients are unequal in the opposite order.

(If $a > b$, c and d are negative and $c = d$, then $a \div c < b \div d$.)

Postulate 13

Given real numbers a and b, exactly one of the following is true: $a < b$, $a = b$, or $a > b$. This is commonly referred to as the Uniqueness of Order Postulate, or the Trichotomy Postulate.

Theorem 1

The shortest segment joining a line and a point outside the line is the perpendicular segment from the point to the line.

Theorem 2

For any real numbers a, b, and c, if $c = a + b$ and $a > 0$, then $c > b$.

7.2 Inequalities in Triangles

Theorem 1

The sum of the lengths of two sides of a triangle is greater than the length of the third side.

Theorem 2

The measures of an exterior angle of a triangle is greater than the measure of either nonadjacent interior angle.

Theorem 3

If two sides of a triangle are unequal, the angles opposite these sides are unequal and the greater angle lies opposite the greater side.

Theorem 4

If two angles of a triangle are unequal, the sides opposite these angles are unequal and the greater side lies opposite the greater angle.

Theorem 5

If two sides of one triangle are equal to two sides of a second triangle and the included angle of the first is greater than the included angle of the second, then the third side of the first triangle is greater than the third side of the second.

Theorem 6

If two sides of one triangle are equal to two sides of a second triangle and the third side of the first is greater than the third side of the second, then the angle opposite the third side of the first triangle is greater than the angle opposite the third side of the second.

7.3 Inequalities in Circles

Theorem 1

In the same circle or in equal circles, if two central angles are unequal, then their arcs are unequal, and the greater angle has the greater arc.

Theorem 2

In the same circle, or in equal circles, if two arcs are unequal, then their central angles are unequal, and the greater arc has the greater central angle.

Theorem 3

In the same circle, or in equal circles, if two chords are unequal, then they are at unequal distances from the center. The longer a chord is, the smaller is the distance from the center.

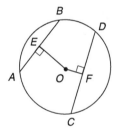

Figure 7.1

If $CD > AB$, $\overline{OE} \perp \overline{AB}$, and $\overline{OF} \perp \overline{CD}$, then $\overline{OF} < \overline{OE}$

CHAPTER 8

Geometric Proportions and Similarity

8.1   Ratio and Proportion

Definition 1

A ratio is the comparison of one number to another, expressed in quotient form. A ratio is therefore a fraction, with a denominator not equal to zero. The ratio of a to b ($b \neq 0$) is represented by a/b.

Definition 2

A proportion is a statement that equates two ratios.

Definition 3

In the proportion $a/b = c/d$, the numbers a and d are called the extremes of the proportion, and the numbers b and c are called the means of the proportion. The single term d is called the fourth proportional.

Theorem 1

In a proportion, the product of the means is equal to the product of the extremes.

(If $a/b = c/d$, then $bc = ad$.)

Theorem 2

A proportion may be written by inversion.

(If $a/b = c/d$, then $b/a = d/c$.)

Theorem 3

The means may be interchanged in any proportion.

(If $a/b = c/d$, then $a/c = b/d$.)

Theorem 4

The extremes may be interchanged in any proportion.

(If $a/b = c/d$, then $d/b = c/a$.)

Theorem 5

A proportion may be written by addition.

(If $\dfrac{a}{b} = \dfrac{c}{d}$, then $\dfrac{a+b}{b} = \dfrac{c+d}{d}$.)

Theorem 6

A proportion may be written by subtraction.

(If $\dfrac{a}{b} = \dfrac{c}{d}$, then $\dfrac{a-b}{b} = \dfrac{c-d}{d}$.)

Theorem 7

If three terms of one proportion are equal, respectively, to three terms of a second proportion, the fourth terms are equal.

(If $\dfrac{a}{b} = \dfrac{c}{d}$, and $\dfrac{a}{b} = \dfrac{c}{e}$, then $d = e$.)

Theorem 8

If the numerators of a proportion are equal, then the denominators are equal.

(If $a/b = c/d$ and $a = c$, then $b = d$.)

Theorem 9

Given a proportion, the ratio of the sum of the numerators to the sum of the denominators forms a proportion with either of the original ratios.

(If $\dfrac{a}{b} = \dfrac{c}{d}$, then $\dfrac{a+c}{b+d} = \dfrac{a}{b}$, and $\dfrac{a+c}{b+d} = \dfrac{c}{d}$.)

Theorem 10

If the product of two numbers (not zero) is equal to the product of two other numbers (not zero), either pair of numbers may be made the means and the other pair may be made the extremes in a proportion.

(If $ab = cd \neq 0$, then $a/c = d/b$ and $b/c = d/a$.

Also, $c/a = b/d$ and $d/a = b/c$.)

8.2   Parallel Lines and Proportions

Theorem 1

A line parallel to one side of a triangle divides the other two sides proportionally.

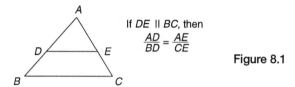

If $DE \parallel BC$, then
$$\frac{AD}{BD} = \frac{AE}{CE}$$

Figure 8.1

Theorem 2

If a line divides two sides of a triangle proportionally, it is parallel to the third side.

Theorem 3

The bisector of one angle of a triangle divides the opposite side in the same ratio as the other two sides.

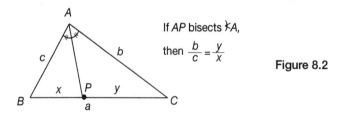

If AP bisects $\angle A$,

then $\dfrac{b}{c} = \dfrac{y}{x}$

Figure 8.2

Theorem 4

Three or more parallel lines intercept proportional segments on any two transversals.

71

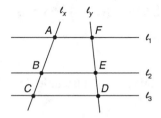

Figure 8.3

If $\ell_1 \parallel \ell_2 \parallel \ell_3$, and ℓ_x and ℓ_y are transversals intersecting the parallel lines at points A through F as shown in figure 8.3, then $\dfrac{AB}{BC} = \dfrac{FE}{DE}$

8.3   Similar Triangles

Definition 1

Two polygons are similar if their corresponding angles are equal and their corresponding sides are in proportion. The symbol for similarity is ~; thus, $\triangle ABC$ is similar to $\triangle DEF$ is written $\triangle ABC \sim \triangle DEF$.

Definition 2

The ratio of similitude refers to the common ratio of corresponding sides of similar polygons.

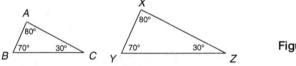

Figure 8.4

$\triangle ABC \sim \triangle XYZ$; $\angle A = \angle X$, $\angle B = \angle Y$, $\angle C = \angle Z$ as shown,

therefore $\dfrac{AB}{XY} = \dfrac{AC}{XZ} = \dfrac{BC}{YZ}$

Postulate

Two triangles are similar if and only if two angles of one triangle are equal to two angles of the other triangle.

72

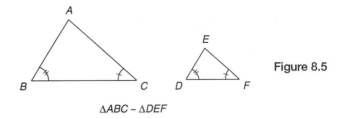

Figure 8.5

△ABC ~ △DEF

Theorem 1

All congruent triangles are similar.

Theorem 2

Two triangles similar to the same triangle are similar to each other.

Theorem 3

If one triangle has an angle equal to that of another triangle, and the respective sides including these angles are proportionate, the triangles are similar.

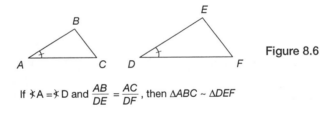

Figure 8.6

If ⊀A = ⊀D and $\frac{AB}{DE} = \frac{AC}{DF}$, then △ABC ~ △DEF

Theorem 4

If three sides of one triangle are in proportion to the three corresponding sides of a second triangle, the triangles are similar (S.S.S. Similarity Theorem).

Theorem 5

If three angles of one triangle are congruent to three corresponding angles of another triangle, then the two triangles are similar (A.A.A. Similarity Theorem).

Theorem 6

If two triangles are similar, the measures of corresponding altitudes have the same ratio as the measures of any two corresponding sides of the triangles.

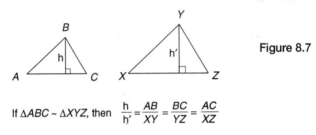

If $\triangle ABC \sim \triangle XYZ$, then $\dfrac{h}{h'} = \dfrac{AB}{XY} = \dfrac{BC}{YZ} = \dfrac{AC}{XZ}$

Figure 8.7

Theorem 7

The perimeters of two similar triangles have the same ratio as the measures of any pair of corresponding sides of the triangles.

Theorem 8

If a given triangle is similar to a triangle that is congruent to a third triangle, then the given triangle is similar to the third triangle.

Theorem 9

Given $\triangle ABC$, if $\dfrac{AB}{AD} = \dfrac{AC}{AE}$, then $\dfrac{AB}{AD} = \dfrac{BC}{DE}$.

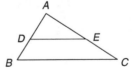

Figure 8.8

Theorem 10

The ratio of similitude of any pair of similar triangles equals the square root of the ratio of their areas.

Theorem 11

The ratio of the areas of any two similar triangles is equal to the ratio of the squares of the lengths of any two corresponding sides, or any two corresponding line segments of the two similar triangles.

Corollary 1

If two triangles are similar, then their corresponding sides are in proportion.

Corollary 2

The measures of any two corresponding line segments of two similar triangles have the same ratio as the measures of any pair of corresponding sides.

Corollary 3

If a line parallel to one side of a triangle intersects the other two sides, then it cuts off a triangle similar to the original triangle.

Corollary 4

Two triangles which are similar to the same triangle are similar to each other. If $\triangle ABC \sim \triangle DEF$, and $\triangle GHI \sim \triangle DEF$, then $\triangle ABC \sim \triangle GHI$.

Corollary 5

If the corresponding sides of two triangles are parallel to each other, the triangles are similar.

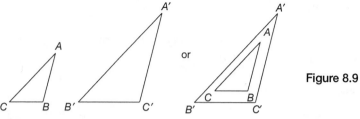

Figure 8.9

In both cases, *AB* ‖ *A′B′*, *BC* ‖ *B ′C′*, and *CA* ‖ *C′A′*; hence, △*ABC* ~ △*A′B′C′*

Corollary 6

If the corresponding sides of two triangles are perpendicular to each other, the triangles are similar.

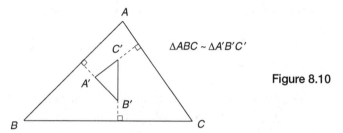

△*ABC* ~ △*A′B′C′*

Figure 8.10

8.4 Properties of the Right Triangle and Similar Right Triangles

Theorem 1

The altitude on the hypotenuse of a right triangle is the mean proportional between the segments of the hypotenuse.

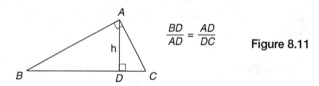

$$\frac{BD}{AD} = \frac{AD}{DC}$$

Figure 8.11

Theorem 2

In a right triangle $\triangle ABC$, the altitude on the hypotenuse separates the triangle into two triangles that are similar to each other and to the original triangle.

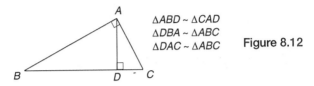

$\triangle ABD \sim \triangle CAD$
$\triangle DBA \sim \triangle ABC$
$\triangle DAC \sim \triangle ABC$

Figure 8.12

Theorem 3

Two right triangles are similar if the ratios of their hypotenuses and any pair of corresponding sides are proportional.

In $\triangle ABC$ and $\triangle CDE$ sharing right angle C, if $\dfrac{AC}{EC} = \dfrac{AB}{ED}$, then $\triangle ABC \sim \triangle EDC$ and $\dfrac{AC}{EC} = \dfrac{BC}{DC}$.

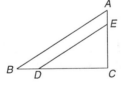

Figure 8.13

Theorem 4

The length of each leg of a given right triangle is the mean proportional between the length of the whole hypotenuse and the length of the projection of that leg on the hypotenuse.

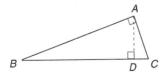

Figure 8.14

$$\frac{BC}{AB} = \frac{AB}{BD}, \frac{BC}{AC} = \frac{AC}{DC}$$

Corollary

If acute angles of two right triangles are congruent, then the triangles are similar.

8.5 Trigonometric Ratios

For definitions 1–4, refer to figure 8.15. Given the right triangle $\triangle ABC$:

Figure 8.15

Definition 1

$$\sin A = \frac{BC}{AB}$$

$$= \frac{\text{measure of side opposite } \measuredangle\ A}{\text{measure of hypotenuse}}$$

Definition 2

$$\cos A = \frac{AC}{AB}$$

$$= \frac{\text{measure of side adjacent to } \measuredangle\ A}{\text{measure of hypotenuse}}$$

Definition 3

$$\tan A = \frac{BC}{AC}$$

$$= \frac{\text{measure of side opposite } \not{\times} A}{\text{measure of side adjacent to } \not{\times} A}$$

Definition 4

$$\cot A = \frac{AC}{BC}$$

$$= \frac{\text{measure of side adjacent to } \not{\times} A}{\text{measure of side opposite } \not{\times} A}$$

Hint: To remember the ratios for sine, cosine and tangent, think of the acronym SOHCAHTOA.

S O H	C A H	T O A
sine = opposite ÷ hypotenuse	cosine = adjacent ÷ hypotenuse	tangent = opposite ÷ adjacent

cotangent is merely the inverse of tangent.

α	$\sin \alpha$	$\cos \alpha$	$\tan \alpha$	$\cot \alpha$
0°	0	1	0	undefined
30°	$\dfrac{1}{2}$	$\dfrac{\sqrt{3}}{2}$	$\dfrac{1}{\sqrt{3}} = \dfrac{\sqrt{3}}{3}$	$\sqrt{3}$
45°	$\dfrac{1}{\sqrt{2}} = \dfrac{\sqrt{2}}{2}$	$\dfrac{1}{\sqrt{2}} = \dfrac{\sqrt{2}}{2}$	1	1
60°	$\dfrac{\sqrt{3}}{2}$	$\dfrac{1}{2}$	$\sqrt{3}$	$\dfrac{1}{\sqrt{3}} = \dfrac{\sqrt{3}}{3}$
90°	1	0	undefined	0

Hint: For the ratios of 30° and 60° think of the 1, 2, $\sqrt{3}$ right triangle:

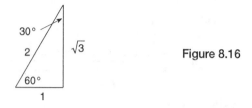

Figure 8.16

For the ratios of 45° think of the 1, 1, $\sqrt{2}$ right triangle:

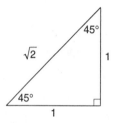

Figure 8.17

Hence, all of the ratios follow.

8.6   Similar Polygons

Definition

Two polygons are similar if there is a one-to-one correspondence between their vertices such that all pairs of corresponding angles are congruent and the ratios of the measures of all pairs of corresponding sides are equal.

Theorem 1

The perimeters of two similar polygons have the same ratio as the measure of any pair of corresponding line segments of the polygons.

Theorem 2

The ratio of the lengths of two corresponding diagonals of two similar polygons is equal to the ratio of the lengths of any two corresponding sides of the polygons.

Theorem 3

The perimeters of two similar polygons have the same ratio as the measures of any pair of corresponding sides of the polygons.

Theorem 4

Two polygons composed of the same number of triangles similar each to each, and similarly placed, are similar.

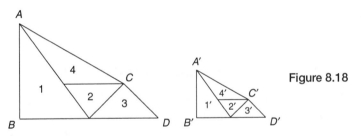

Figure 8.18

Properties of Similarity

Reflexive property

Polygon *ABCD* ~ Polygon *ABCD*.

Symmetric property

If polygon *ABCD* ~ polygon *EFGH*, then polygon *EFGH* ~ polygon *ABCD*.

Transitive Property

If polygon *ABCD* ~ polygon *EFGH*, and polygon *EFGH* ~ polygon *WXYZ*, then polygon *ABCD* ~ polygon *WXYZ*.

Summary of Essential Definitions, Theorems, and Properties

Points, Lines, Planes, and Angles

If A and B are two points on a line, then the line segment AB is the set of points on that line between A and B and including A and B, which are called the endpoints. The line segment is referred to as $\overline{AB}$.

An angle is a collection of points which is the union of two rays having the same endpoint. An angle such as the one illustrated can be referred to in any of the following ways:

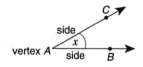

A) by a capital letter which names its vertex, i.e., ∠A;

B) by a lowercase letter or number placed inside the angle, i.e., ∠x;

C) by three capital letters, where the middle letter is the vertex and the other two letters are not on the same ray, i.e., ∠CAB or ∠BAC, both of which represent the angle illustrated in the figure on the previous page.

Angle Sum

If A is in the interior of ∠XYZ, then $m \angle XYZ = m \angle XYA + m \angle AYZ$.

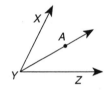

Angle Difference

If P is in the exterior of ∠ABC and in the same half-plane (created by edge $\overleftrightarrow{BC}$) as A, then $m \angle ABP = m \angle PBC - m \angle ABC$.

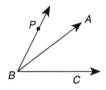

A ray bisects (is the bisector of) an angle if the ray divides the angle into two angles that have equal measure. $m \angle BAD = m \angle DAC$

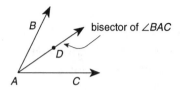

If the two non-common sides of adjacent angles form opposite rays, then the angles are called a linear pair. Note that $\angle \alpha$ and $\angle \beta$ are supplementary.

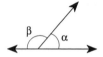

Two lines are said to be perpendicular if they intersect and form right angles. The symbol for perpendicular (or, is therefore perpendicular to) is $\perp$; $\overleftrightarrow{AB}$ is perpendicular to $\overleftrightarrow{CD}$ is written $\overleftrightarrow{AB} \perp \overleftrightarrow{CD}$.

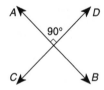

If two lines intersect and form one right angle, then the lines form four right angles.

Vertical angles are equal.

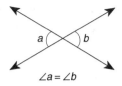

$\angle a = \angle b$

Two supplementary angles are right angles if they have the same measure.

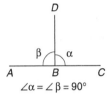

$\angle \alpha = \angle \beta = 90°$

Two intersecting lines are perpendicular to each other if they form right angles. $\overleftrightarrow{AB} \perp \overleftrightarrow{CD}$.

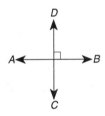

Adjacent angles are supplementary if their exterior sides form a straight line. $\angle \alpha + \angle \beta = 180°$.

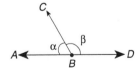

If the exterior sides of adjacent angles are perpendicular to each other, then the adjacent angles are complementary. $\angle\alpha + \angle\beta = 90°$.

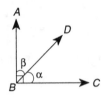

If $m \angle ABC = m \angle DEF$, and $\overrightarrow{BX}$ and $\overrightarrow{EY}$ bisect $\angle ABC$ and $\angle DEF$, respectively, then $m \angle ABX = m \angle DEY$.

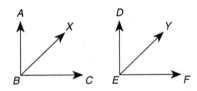

Triangles

A triangle with a right angle is called a right triangle. The side opposite the right angle in a right triangle is called the hypotenuse. The other two sides are called the legs of the right triangle. AC is the hypotenuse.

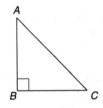

An altitude of a triangle is a line segment from a vertex of the triangle perpendicular to the opposite side.

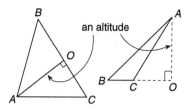

A line segment connecting a vertex of a triangle and the midpoint of the opposite side is called a median of the triangle.

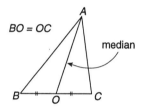

A line that bisects and is perpendicular to a side of a triangle is called a perpendicular bisector of that side.

An angle bisector of a triangle is a line that bisects an angle and extends to the opposite side of the triangle.

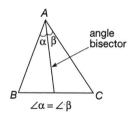

The line segment that joins the midpoints of two sides of a triangle is called a midline of the triangle.

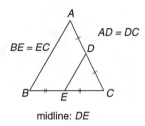

midline: *DE*

An exterior angle of a triangle is an angle formed outside a triangle by one side of the triangle and the extension of an adjacent side. α is an exterior angle.

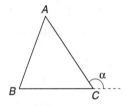

Inequalities in Triangles

The sum of the lengths of two sides of a triangle is greater than the length of the third side.

The measure of an exterior angle of a triangle is greater than the measure of either nonadjacent interior angle.

If two sides of a triangle are unequal, the angles opposite these sides are unequal and the greater angle lies opposite the greater side.

Congruent Angles

Two triangles are congruent if there is a one-to-one correspondence between their vertices such that all pairs of corresponding sides have equal measures and all pairs of corresponding angles have equal measures. This is denoted by $\cong$.

Areas of Triangles

The area of a triangle is given by the formula $A = \dfrac{1}{2}bh$, where b is the length of a base and h is the corresponding height of the triangle.

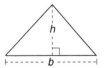

Parallelism

A transversal of two or more parallel lines is a line that cuts across these lines in two or more points, one point for each line.

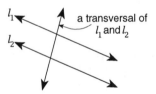

a transversal of l_1 and l_2

If two parallel lines are cut by a transversal, nonadjacent angles on opposite sides of the transversal but on the interior of the two lines are called alternate interior angles and are equal. $\angle\alpha = \angle\beta$.

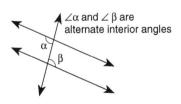

$\angle\alpha$ and $\angle\beta$ are alternate interior angles

89

If two parallel lines are cut by a transversal, nonadjacent angles on opposite sides of the transversal and on the exterior of the two lines are called alternate exterior angles and they are equal. $\angle\alpha = \angle\beta$.

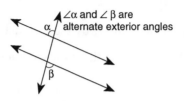

$\angle\alpha$ and $\angle\beta$ are alternate exterior angles

If two parallel lines are cut by a transversal, angles on the same side of the transversal and in corresponding positions with respect to the lines are called corresponding angles. $\angle\alpha = \angle\beta$.

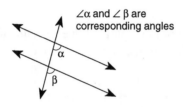

$\angle\alpha$ and $\angle\beta$ are corresponding angles

If a line is perpendicular to one of two parallel lines, then it is perpendicular to the other line, too.

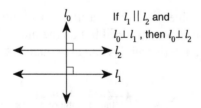

If $l_1 \parallel l_2$ and $l_0 \perp l_1$, then $l_0 \perp l_2$

If two lines being cut by a transversal form congruent corresponding angles, then the two lines are parallel.

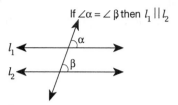

If $\angle\alpha = \angle\beta$ then $l_1 \parallel l_2$

Quadrilaterals

A quadrilateral is a polygon with four sides.

A parallelogram is a quadrilateral whose opposite sides are parallel. $AC \parallel BD$ and $AB \parallel CD$.

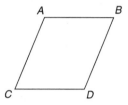

A rectangle is a parallelogram with one right angle. Therefore, all angles are right angles.

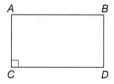

Opposite sides of a parallelogram are equal.

Nonconsecutive angles of a parallelogram are equal.

If both pairs of opposite angles of a quadrilateral are congruent, the quadrilateral is a parallelogram.

If one angle of a quadrilateral is congruent to the opposite angle and one side is parallel to the opposite side, then the quadrilateral is a parallelogram.

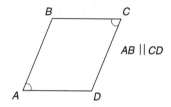

$AB \parallel CD$

If one angle of a quadrilateral is congruent to its opposite, and one side is congruent to its opposite, then the quadrilateral is a parallelogram.

Segments of parallel lines intercepted between parallel lines are congruent.

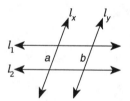

If $l_1 \parallel l_2$ and $l_x \parallel l_y$, then $a = b$.

A rhombus is a parallelogram with two adjacent sides equal. $AB \cong BD$.

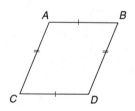

All sides of a rhombus are equal.

A square is a rhombus with a right angle.

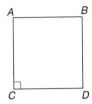

A trapezoid is a quadrilateral with two and only two sides parallel. The parallel sides of a trapezoid are called bases.

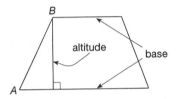

Similarity

The bisector of one angle of a triangle divides the opposite side in the same ratio as the other two sides. $\dfrac{x}{y} = \dfrac{c}{b}$.

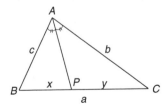

The ratio of similitude refers to the common ratio of corresponding sides of similar polygons. $\dfrac{AB}{BC} = \dfrac{DE}{EF}$.

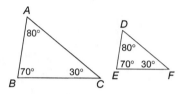

Two triangles are similar if, and only if, two angles of one triangle are equal to two angles of the other triangle.

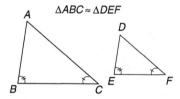

$\triangle ABC \approx \triangle DEF$

The altitude on the hypotenuse of a right triangle is the mean proportional between the segments of the hypotenuse.

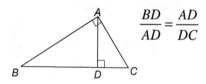

$$\dfrac{BD}{AD} = \dfrac{AD}{DC}$$

Trigonometric Ratios

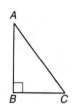

$$\sin A = \frac{\text{measure of side opposite } \angle A}{\text{measure of hypotenuse}}$$

$$\cos A = \frac{\text{measure of side adjacent } \angle A}{\text{measure of hypotenuse}}$$

$$\tan A = \frac{\text{measure of side opposite } \angle A}{\text{measure of side adjacent } \angle A}$$

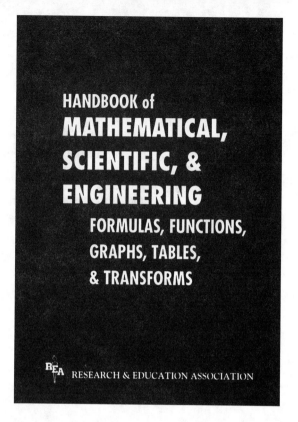

HANDBOOK of MATHEMATICAL, SCIENTIFIC, & ENGINEERING FORMULAS, FUNCTIONS, GRAPHS, TABLES, & TRANSFORMS

RESEARCH & EDUCATION ASSOCIATION

A particularly useful reference for those in math, science, engineering and other technical fields. Includes the most-often used formulas, tables, transforms, functions, and graphs which are needed as tools in solving problems. The entire field of special functions is also covered. A large amount of scientific data which is often of interest to scientists and engineers has been included.

Available at your local bookstore or order directly from us by sending in coupon below.

RESEARCH & EDUCATION ASSOCIATION
61 Ethel Road W., Piscataway, New Jersey 08854
Phone: (732) 819-8880   website: www.rea.com

VISA   MasterCard

☐ Payment enclosed
☐ Visa   ☐ MasterCard

Charge Card Number

Expiration Date: _____ / _____
                    Mo   Yr

Please ship the **"Math Handbook"** @ $34.95 plus $4.00 for shipping.

Name _____

Address _____

City _____ State _____ Zip _____

REA's **Problem Solvers**

The "PROBLEM SOLVERS" are comprehensive supplemental text-books designed to save time in finding solutions to problems. Each "PROBLEM SOLVER" is the first of its kind ever produced in its field. It is the product of a massive effort to illustrate almost any imaginable problem in exceptional depth, detail, and clarity. Each problem is worked out in detail with a step-by-step solution, and the problems are arranged in order of complexity from elementary to advanced. Each book is fully indexed for locating problems rapidly.

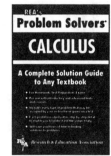

ACCOUNTING
ADVANCED CALCULUS
ALGEBRA & TRIGONOMETRY
AUTOMATIC CONTROL
  SYSTEMS/ROBOTICS
BIOLOGY
BUSINESS, ACCOUNTING, & FINANCE
CALCULUS
CHEMISTRY
COMPLEX VARIABLES
DIFFERENTIAL EQUATIONS
ECONOMICS
ELECTRICAL MACHINES
ELECTRIC CIRCUITS
ELECTROMAGNETICS
ELECTRONIC COMMUNICATIONS
ELECTRONICS
FINITE & DISCRETE MATH
FLUID MECHANICS/DYNAMICS
GENETICS
GEOMETRY
HEAT TRANSFER

LINEAR ALGEBRA
MACHINE DESIGN
MATHEMATICS for ENGINEERS
MECHANICS
NUMERICAL ANALYSIS
OPERATIONS RESEARCH
OPTICS
ORGANIC CHEMISTRY
PHYSICAL CHEMISTRY
PHYSICS
PRE-CALCULUS
PROBABILITY
PSYCHOLOGY
STATISTICS
STRENGTH OF MATERIALS &
  MECHANICS OF SOLIDS
TECHNICAL DESIGN GRAPHICS
THERMODYNAMICS
TOPOLOGY
TRANSPORT PHENOMENA
VECTOR ANALYSIS

If you would like more information about any of these books,
complete the coupon below and return it to us or visit your local bookstore.

RESEARCH & EDUCATION ASSOCIATION
61 Ethel Road W. • Piscataway, New Jersey 08854
Phone: (732) 819-8880   **website: www.rea.com**

Please send me more information about your Problem Solver books

Name _____

Address _____

City _____ State _____ Zip _____

REA's Test Preps
The Best in Test Preparation

- REA "Test Preps" are **far more** comprehensive than any other test preparation series
- Each book contains up to **eight** full-length practice tests based on the most recent exams
- **Every** type of question likely to be given on the exams is included
- Answers are accompanied by **full** and **detailed** explanations

REA publishes over 70 Test Preparation volumes in several series. They include:

Advanced Placement Exams (APs)
Biology
Calculus AB & Calculus BC
Chemistry
Economics
English Language & Composition
English Literature & Composition
European History
Government & Politics
Physics B & C
Psychology
Spanish Language
Statistics
United States History

College-Level Examination Program (CLEP)
Analyzing and Interpreting Literature
College Algebra
Freshman College Composition
General Examinations
General Examinations Review
History of the United States I
History of the United States II
Human Growth and Development
Introductory Sociology
Principles of Marketing
Spanish

SAT Subject Tests
Biology E/M
Chemistry
English Language Proficiency Test
French
German

SAT Subject Tests (cont'd)
Literature
Mathematics Level 1, 2
Physics
Spanish
United States History
Writing

Graduate Record Exams (GREs)
Biology
Chemistry
Computer Science
General
Literature in English
Mathematics
Physics
Psychology

ACT - ACT Assessment

ASVAB - Armed Services Vocational Aptitude Battery

CBEST - California Basic Educational Skills Test

CDL - Commercial Driver License Exam

CLAST - College Level Academic Skills Test

COOP & HSPT - Catholic High School Admission Tests

ELM - California State University Entry Level Mathematics Exam

FE (EIT) - Fundamentals of Engineering Exams - For both AM & PM Exams

FTCE - Florida Teacher Certification Exam

GED - High School Equivalency Diploma Exam (U.S. & Canadian editions)

GMAT CAT - Graduate Management Admission Test

LSAT - Law School Admission Test

MAT - Miller Analogies Test

MCAT - Medical College Admission Test

MTEL - Massachusetts Tests for Educational Licensure

NJ HSPA - New Jersey High School Proficiency Assessment

NYSTCE: LAST & ATS-W - New York State Teacher Certification

PLT - Principles of Learning & Teaching Tests

PPST - Pre-Professional Skills Tests

PSAT - Preliminary Scholastic Assessment Test

SAT

TExES - Texas Examinations of Educator Standards

THEA - Texas Higher Education Assessment

TOEFL - Test of English as a Foreign Language

TOEIC - Test of English for International Communication

USMLE Steps 1,2,3 - U.S. Medical Licensing Exams

U.S. Postal Exams 460 & 470

RESEARCH & EDUCATION ASSOCIATION
61 Ethel Road W. • Piscataway, New Jersey 08854
Phone: (732) 819-8880   **website: www.rea.com**

Please send me more information about your Test Prep books

Name _____

Address _____

City _____ State _____ Zip _____